I0484019

CLIMATE CHANGE and VIETNAM

The darkness palpitated down upon all this, and then the real thing came at last. It was something formidable and swift, like the sudden smashing of a Vial of Wrath. It seemed to explode all round the ship with an overpowering concussion and a rush of great waters, as if an immense dam had been blown up to windward.
- Joseph Conrad, *Typhoon*[1]

Introduction

As sentinels to disaster, the first to perish under the onslaught of Typhoon Linda in 1997 were Vietnamese sailors and fisherman caught unawares out to sea.[2] They would soon be followed by thousands of their countrymen; and by the time Linda passed over Vietnam and into the Gulf of Thailand, she would leave behind 3,111 dead and $385 million USD in damage.[3] Like those sailors, Vietnam is likely to be one of the nations first in vulnerability to the progress of global climate change - a distinction it shares with regional neighbors like Bangladesh, India and other South-East Asian nations characterized by: densely populated, extensive coastlines; exposure to typhoon/cyclone patterns; environmental degradation; and developing world technology and infrastructure.[4] That list of common characteristics highlights two themes at the core of this paper. First, these vulnerabilities cross a number of economic, social and physical sectors; and so addressing the potential associated consequences demands a similarly cross-cutting approach by a diverse set of actors,

---

[1] Joseph Conrad, *Typhoon* (New York: Putnam, 1902), 76-77.

[2] C.P. Dillon and Mark Andrews. *1997 Annual Tropical Cyclone Report* (Pearl Harbor, HI: Joint Typhoon Warning Center, 1997), 116.

[3] UN Development Program, "Summing-up report on disaster situations in recent years and preparedness and mitigation measures in Vietnam," *Reliefweb.int*, 24 February 2003, accessed 5 October 2013, http://reliefweb.int/report/viet-nam/summing-report-disaster-situations-recent-years-and-preparedness-and-mitigation.

[4] Ole Bruun and Thorkill Casse, "Climate Change, Adaptation and the Environment in Central Vietnam," in *On the Frontiers of Climate and Environmental Change: Vulnerabilities and Adaptations in Central Vietnam,* ed. Ole Bruun et al. (New York: Springer, 2013), 2.

including the Government of Vietnam, Non-Governmental/Intergovernmental Organizations (NGO/IGO) and the assistance efforts of other nations. Second, while these vulnerabilities are distinctly Vietnamese, they are not uniquely so; and thus efforts taken to address them in a Vietnamese context may prove useful in countries that share these vulnerabilities – either as program models to be replicated, or as program elements to be incorporated in a broader, regional approach.

This paper will explore these two themes by initially reviewing key underlying characteristics of the Vietnamese milieu most relevant to climate change, and considering specifically how climate change is likely to interact with these characteristics. I will then discuss the critical players working to address these issues, both within and outside the Government of Vietnam, with particular attention to those elements that might most benefit from additional outside assistance. After briefly reviewing possible counter-arguments, the study concludes with proposing potential forms of such assistance. Although specific recommendations for U.S. Pacific Command lie outside the remit of this paper, the analysis outlined here will be scoped and focused so that such recommendations might reasonably flow from subsequent analysis.

Ultimately, this paper will demonstrate that Vietnam is indeed critically vulnerable to climate change, and that assistance in this sector is an appropriate avenue for expanded U.S. support; but first, the discussion begins by returning to Typhoon Linda, where bodies of the victims washed up on Vietnamese shores weeks after the event.[5]

_____

[5] Reuters, "Typhoon victims' bodies found on Vietnam shores," *The Nation* 15 November 1997, accessed 5 October 2013, http://news.google.com.

Impacts

While perhaps lurid, a typhoon-related vignette is an entirely appropriate introduction of Vietnamese vulnerabilities to the effects of climate change. Given Vietnam's geographic location, these devastating weather events have been a permanent feature of Vietnamese life (prompting mitigation efforts as early as the First Century, C.E.); and in more recent times, have often struck up to ten times a year, impacting roughly 60% of the population.[6] Notably, the frequency of the most severe storms (12 or greater on the Beaufort scale) has apparently begun to increase: only two such storms struck Vietnam in each of the decades 1981-90 and 1991-2000; while in the years 2001-08, the number jumped to five.[7] With almost all of its 3,260 km of coastline (including population centers like Ho Chi Minh City) exposed to regional cyclone tracks, Vietnam will be vulnerable to any change in storm activity.[8] To date, primarily due to limited regional historical observation outside the North Atlantic (which is likely to see an increase in frequency of the most powerful storms), there is little current data available to model increased intensity or frequency of these storms in South East Asia – however, data does exist to suggest a likely increase in extreme precipitation in storms making landfall across the region.[9] Highlighting the complex inter-system dynamics associated with climate change, it is just such an increase in extreme precipitation, in conjunction with increased vulnerability to storm surges driven by sea level rise (rather than the force of the storm itself), that is responsible more much of the cyclone-associated

---

[6] Dang Quang Tinh, "Viet Nam – Country Report," *Asian Disaster Reduction Center*, 1999, accessed 8 October 2013, http://www.adrc.asia/countryreport/VNM.

[7] Bruun and Casse, *On the Frontiers of Climate and Environmental Change*, 3.

[8] Michael Waibel, "Implications and Challenges of Climate Change for Vietnam," *Pacific News*, January-February 2008, accessed 15 September 2013, http://www.pacific-geographies.org.

[9] Intergovernmental Panel on Climate Change (IPCC), "Climate Change 2013: The Physical Science Basis," *IPCC*, September 2013, 69-71, accessed 10 October 2013, http://www.ipcc.ch.

damage.[10] In addition to their role in exacerbating the impact of cyclones, changes in precipitation and sea level will have a variety of other impacts that merit independent consideration.

Based on mid-range IPCC scenario forecasts, Vietnam is likely to experience a sea level rise of 75cm by 2100. In the same scenario, precipitation will increase by 5% annually, and average temperatures will increase by 2.3 C. Critically, the rise in these latter two factors will not be uniform across the year – rather, precipitation will rise significantly in the traditional rainy season and decrease in the dry season, while temperatures will rise more rapidly in winter than in summer.[11] That change in relative variability of temperature and rainfall, in conjunction with the absolute rise, is likely to: increase the frequency of both droughts and flooding, increase the danger of forest fires, and threaten the extinction of various plants and animals (including economically important timber plants). Rising sea level, in turn, is likely to drive increased salinity of fresh water resources, threatening: significant cultivation areas in the Mekong and North Deltas, important aquaculture species, and the area and depth of coastal mangrove forests.[12] Reinforcing the cyclical nature of these issues, the very mangrove forests threatened by increased salinity are an important natural bulwark stabilizing shorelines against storms and wave damage.[13] Economic consequences

---

[10] IPCC, *Managing the Risks of Extreme Events and Disasters to Advance Climate Change Adaptation* (New York: Cambridge, 2012), 158.

[11] Ministry of Natural Resources and Environment, Government of Vietnam, "Climate Change, Sea Level Rise Scenarios for Vietnam," *Preventionweb*, June 2008, accessed 10 September 2013, http://www.preventionweb.net

[12] Institute of Strategy and Policy on Natural Resources and Environment, Government of Vietnam, "Viet Nam Assessment Report on Climate Change," *United Nations Environment Program*, 2009, 75-76, accessed 10 September 2013, http://www.unep.org.

[13] Eleanor Jane Sterling, Martha Maud Hurley and Le Duc Minh, *Vietnam: A Natural History* (New Haven: Yale University Press, 2006), 91-92.

of these changes could include an aggregate decline in GDP of 2.3-2.4% by 2050 based on agricultural impacts, and deterioration of infrastructure across Vietnam's 116 ports.[14]

While these macro-level impacts will be felt broadly across Vietnam (where 76% of the population relies upon the affected natural resource base in one way or another),[15]variations in demographic, economic and physical landscapes will lead to variations in precisely how these impacts are faced. For example, the rural poor may be more vulnerable in an immediate, practical sense to the impact of both natural disaster and economic disruption; but in a relative sense, they have less to lose than the wealthy in terms of capital and physical plant in the face of the same threats. Alternatively, vulnerability can be framed in terms both of immediate exposure to threat (external forces) and ability to cope with the consequences of a threat (internal forces).[16] Mapped this way, the coastal regions and urban centers are more exposed to risks like cyclones and sea level rise than the less densely populated, rugged hinterlands; but the former, by virtue of economic development, have greater internal resources for consequence management relative to the latter.

Complicating matters further, actions taken by the government to promote economic development can have unintended consequences for this vulnerability mapping in a variety of ways. Privatization of land and encouragement of growing revenue-generating crops for export can clearly serve to promote economic growth benefiting the wealthy; however, that same growth promotes inequality and undermines the coping capacity of the landless poor as

---

[14] World Bank, "Economics of Adaptation to Climate Change, Vietnam," *Climate Change*, World Bank, 2010, xv-xvii, accessed 10 September 2013, http://climatechange.worldbank.org.
[15] W. Adger, P. Mick Kelly and Nguyen Huu Ninh, "Environment, society and precipitous change," in *Living with Environmental Change: Social vulnerability, adaptation and resilience in Vietnam.* ed. W. Adger et al. (New York: Routledge, 2001), 3-6.
[16] P.H. Liotta and Allan W. Shearer, *Gaia's Revenge* (Westport, CT: Praeger, 2007), 49-51.

discussed above.[17] Similarly, the expansion of large hydropower and reservoir construction can increase social resilience through associated economic development. However, the same projects undermine physical resilience by increasing the danger of flooding during heavy precipitation in the rainy season; and decreasing water available for irrigation during the dry season (both trends which climate change is expected to exacerbate).[18]

Government of Vietnam Response

Taken in isolation, the negative potential consequences of climate change would appear to merit government action; but as the discussion above suggests, such action may come at a cost against competing priorities like modernization and economic development. In that context, it bears inquiring how the Government of Vietnam assesses the threat – and such inquiry reveals substantive government concern. On 5 December 2011, the Vietnamese Prime Minister's office issued a national strategy document highlighting how the government views the issue and sets out a programmatic response for addressing it. Notably, while the strategy reviews the types of impacts described above in terms of natural disaster and environmental change (which are comparatively value neutral); it also crosses more controversial terrain by addressing many of the specific modernization efforts that might contribute to, or exacerbate, the phenomenon. In addition to initiatives related to early

---

[17] Tobia Platen-Hallermund and Anton Mikkel Thorsen, "Natural Resource Management Impact on Vulnerability in Relation to Climate Change" in *Frontiers of Climate and Environmental Change,* 174.
[18] Morgens Buch-Hansen, Nguyen Ngoc Khanh, and Ngwyen Hong Anh, "Paradoxes in Adaptation: Economic Growth Differentiation, A Case Study of Mid-Central Vietnam" in *Frontiers of Climate and Environmental Change,* 39.

warning and disaster management, the strategy proposes programs for addressing food and water security, as well as greenhouse gas emission reductions.[19]

Within the government, that national strategy assigns the Ministry of Natural Resources and Environment (MNRE) as the lead executive agent for climate change programs. The precedent for this role was established well prior to announcement of the 2011 national strategy. In 2003, just after Vietnam's accession to the 2002 Kyoto Protocol, MNRE was designated as Vietnam's Clean Development Mechanism National Authority (a designation which highlight's the government's cross-sector approach to the issue).[20] As the lead agency for climate change, MNRE's programs flow naturally in two primary channels – mitigation (measures to reduce the factors which contribute to climate change) and adaptation (measures to respond to likely consequences of climate change). Of note, while mitigation issues may often be associated with industrial emissions, in Vietnam the principal sources of greenhouse gasses (GHG) are found in the agricultural sector (specifically rice cultivation, livestock, agricultural soils and burning of agricultural residue). Consequently, priority government programs are aimed not at phenomenon like alternative industrial power sources; but rather at issues like biogas and nutrition enhancement for dairy cattle.[21] In the parallel line of effort, MNRE's priorities for adaptation programs are agriculture, forestry, coastal zone management and water resources management.[22]

---

[19] Prime Minister's Office, Government of Vietnam, "National Strategy on Climate Change," *Government Portal*, 5 December 2011, accessed 15 September 2013, http://www.chinhphu.vn.

[20] MNRE, "Home page," *Government Portal*, Government of Vietnam, accessed 15 September 2013, http://www.noccop.org.vn.

[21] MNRE, *Vietnam Technology Needs Assessment for Climate Change Mitigation* (Ha Noi: Government of Vietnam, June 2012), 4.

[22] MNRE, *Vietnam Technology Needs Assessment for Climate Change Adaptation* (Ha Noi: Government of Vietnam, June 2012), 2.

Across this broad range of program activities, MNRE's primary implementing institution is the Hydro-Meteorological Service (HMS), which runs the National Center for Hydro-Meteorological Forecasting (NCHF). The NCHF (as the name implies) is divided into two functional arms with specific core competencies – meteorological (cyclone forecasting) and hydrological (flood forecasting). In addition to NCHF's role in deliberate programming, these core competencies make it an integral component of the national disaster response community (a community likely to be progressively challenged by overlapping, reinforcing effects of climate change). At the apex, this community is led by the Central Committee for Flood and Storm Control (CFSCC). This organization is supported by mirror committees at the city and provincial levels, and is chartered to: monitor the state of physical defense (e.g. dykes); manage both government and public information during a storm/flood event; and organize post-event relief. Additionally, these committees are supported by a national Disaster Management Unit (DMU), established in 1994, charged with: computer-based emergency warning and damage reporting; GIS distribution management of relief resources; web-based public information; decision support templates; and training at all levels of government.[23] While in principle the outline above describes an impressive state response framework, important caveats apply in practice. Historically, robust disaster assistance has not been viewed as a formal state obligation (by either the governing or governed classes) and relief has often been ad hoc at best. Moreover, strong centralizing tendencies in Vietnamese bureaucratic culture tend to foster local governments that are primarily simple

---

[23] P. Mick Kelly, Hoang Minh Hien and Tran Viet Lin, "Responding to El Nino and La Nina: Averting Tropical Cyclone Impacts," *Living with Environmental Change,* 167-169.

executors of central government policy, undermining the agility and initiative that can be critical in disaster response.[24]

Other Responders

As described above, Vietnam recognizes the challenges posed by climate change and has begun applying a variety of institutions and programs against the problem. Purely internal resources, however, are unlikely to prove equal to the challenge, and will require outside support. Indeed, an explicit priority of MNRE's efforts is to encourage and mobilize international donors to provide relevant resources and technologies.[25] To whom might Vietnam turn in such a quest? As noted in the introduction, the climate change consequences confronting Vietnam are not simply a national problem. They will clearly have a regional impact, and thus regional organizations are likely to be relevant actors. As the preeminent regional organization, the Association of South East Asian Nations (ASEAN) has issued a variety of statements and position papers calling for coordinated action to respond to climate change. While many of these may amount to little more than the broad exhortations which sometimes characterize Intergovernmental Organizations with limited implementation authority, a number of the ASEAN proposals do focus on specific modeling and monitoring technologies and solicitation of broader international assistance in ways that resonate with Vietnam's own initiatives.[26] Another regional organization with potential interest in climate change is the Shanghai Cooperation Organization (SCO). Although the SCO is more focused

---

[24] Olivier Rubin, "Impediments to Climate-Induced Disaster Management: Evidence from Quang Nam Province, Central Vietnam," *Frontiers of Climate and Environmental Change,* 114.

[25] MNRE, "Viet Nam: Policy development, financial mechanism, technology transfer to respond to climate change," *UNFCC,* 28 November 2012, accessed 20 October 2013, http://unfcc.int.

[26] Raman Letchumanan, "Is there an ASEAN policy on climate change?" *Special Report 004,* January 2010, accessed 15 October 2013, http://www.lse.ac.uk.

toward Central Asia, China is a founding member; and South/Southeast Asian actors like India, Sri Lanka and ASEAN itself hold varying degrees of observer/guest relations with the forum. To date, however, the SCO has shown little interest in climate change as an explicit topic, although it does deal with related issues of disaster management and sustainable development. That lacuna is a striking one, given that the organization's titular city of Shanghai – along with Ho Chi Minh City – is the fourth most vulnerable large city in the region to the consequences of climate change.[27]

The SCO's policy position appears to be driven by Chinese perspectives, which introduces the next category of potential actors – regional nation states. China itself is already facing climate impacts ranging from desertification and loss to the sea of land in the Pearl River Basin, to degradation of major infrastructure projects like railway construction. However, aside from a few isolated academics and limited People's Liberation Army reviews, climate change in Chinese discourse has not emerged as a fundamental national security challenge. Instead, it remains tightly confined within lesser priority agricultural and development fields, limiting China's ability to engage in the emerging global terms of debate.[28] A perhaps unintended consequence of such a position, however, is the possibility that foreign interventions in the climate change realm (which figure more prominently in national security discourse elsewhere) are unlikely to trigger Chinese sensitivities associated with security affairs. In marked contrast to China's relative indifference, another major regional player – India – has taken a far more engaged approach to climate change; and is

---

[27] Rafael Senga, "Natural or Unnatural Disasters: the Relative Vulnerabilities of Southeast Asian Megacities to Climate Change," *SR004*, January 2010, accessed 15 October 2013, http://www.lse.ac.uk.
[28] Duncan Freeman, "The Missing Link: China, Climate Change and National Security," *Asia Paper Vol. 5*, 2010, accessed 21 October 2013, http://www.vub.ac.be.

notably expanding its cooperative ties with Vietnam across a broad range of initiatives.[29]
Similarly, Japan has initiated a multi-year program to assist Vietnam in both climate change
amelioration and disaster management from 2013 through 2016.[30] In the latter category of
disaster management, Japan has generally taken on a more active approach to regional
solutions in the wake of the 2011 earthquake and tsunami, including establishment of an
ASEAN disaster information network through the ASEAN Coordinating Center for
Humanitarian Assistance.[31] Like Japan, Australia has newly identified Climate Change
Assistance as a specific program element of its international support efforts. In Vietnam's
case, such programs supported by AusAID focus on food security, livelihoods, and low
carbon growth options – on the scale of $17.7 million USD.[32]

While AusAID and similar national aid agencies conduct programs of their own, as a
matter of practice, much of their programming is executed by another category of actors:
NGOs and IGOs. While the variety and scale of these organizations is kaleidoscopic, a few
examples can illustrate the breadth of activity. Within the traditional UN development
sphere, the International Fund for Agricultural Development pursues a range of climate
change related projects (the majority of them in South and South East Asia); focusing in
Vietnam on climate change planning and carbon sequestration markets.[33]

Reflecting another approach, the Rockefeller Foundation has funded the Asian Cities
Climate Change Resilience Network (ACCCRN), a regional program including India,

[29] Rajaram Panda, "Vietnam and the World: Focus on the U.S. and India." *Eurasia Review,* 12 August 2013, accessed on 21 October 2013, http://www.eurasiareview.com.
[30] Voice of Vietnam, "Japan provides expertise on climate change mitigation," *VOV Online Newspaper*, 13 September 2013, accessed 25 September 2013, http://english.vov.vn.
[31] Government of Vietnam, "Vietnam-Japan cooperation in natural disaster prevention." *Reliefweb*, 3 October 2013, accessed 15 October 2013, http://www.releiefweb.int.
[32] AusAID, "Vietnam: Climate change assistance," *AusAID*, 13 June 2013, accessed 15 October 2013, http://www.ausaid.gov.au.
[33] IFAD, "Planning for Climate Change in Vietnam," *IFAD*, accessed 15 October 2013, http://www.ifad.org.

Indonesia, Thailand in Vietnam. Within Vietnam specifically, ACCCRN is active in Can Tho, Da Nang and Quy Nhon, with programs focusing on urban development and flood patterns, and improving housing for the urban poor to resist increasingly powerful storms.[34] Continuing the theme of regional approaches to climate change – and highlighting Vietnam's central place in this order – the CGIAR Research Program on Climate Change, Agriculture and Food Security recently opened a hub office in Hanoi to manage related programs (with a focus on impacts to rice production) across South East Asia.[35]

Forms of Assistance Required

As suggested by the discussion above, a wide variety of actors both internal and external to Vietnam are pursuing climate change related projects. As is often the case in the international assistance arena, however, many of these programs may be designed based on the preferences, technical capabilities and domestic political priorities of donors as much as they are by the needs of the recipient country. Having reviewed some of the major international efforts underway, a useful point of reference can be made by reviewing in turn the Government of Vietnam's priorities. As articulated by MNRE, these include: developing the climate change monitoring system with high accuracy digital elevation models; develop flood, disaster risk and climate maps with GIS; and issuing policies related to climate change mitigation and GHG mitigation for prioritized sectors: agriculture, forestry, land use, water resource, energy and transportation.[36]

---

[34] ACCRN, "City Progress Watch," *Rockefeller Foundation*, 22 August 2012, accessed 22 October 2013, http://www.www.rockefellerfundation.org.
[35] CCFAS, "Climate change research hub opens in Hanoi," *CGIAR*, 7 May 2013, accessed 1 October 2013, http://ccafs.cgiar.org/regions/southeast-asia.
[36] MNRE, "Viet Nam: Policy development," accessed 20 Oct. 2013, http://unfcc.int.

For an alternative, relatively independent view of potential new initiatives, the U.S. Forest Service (USFS) conducted an assessment that identified three possible program planks. The first – information, education and communication – focuses on local level awareness raising with regards to storm and flood disaster preparedness, sea level rise, and broader water resource issues. A second plank includes technical capacity building, and data collection and interpretation. Similar to the first plank, the emphasis is on local-level, ground up efforts to counter-balance the bureaucratic centralizing tendencies mentioned earlier, which may sometimes undermine effective, adaptive local response. Regarding data collection generally, it merits remark that (despite the sometimes apparent deluge of climate related data) many of the extant data sources are comparatively limited for specific regions like Vietnam, forcing local planners to "down-scope" from global models, introducing even greater uncertainty into already probabilistic forecasts. A third component involves local implementation of pilot projects, including expansion of an existing biogas project and improvement of sea/river dykes and drainage systems.[37]

U.S. Assistance

The provenance of that assessment from the USFS serves as a suitable introduction to the distinctive architecture of U.S. assistance to Vietnam in the climate change realm. More precisely, why ever is a subordinate agency of the Department of the Interior involved in foreign assistance programming? Such programming is traditionally a charter of the Department of State's USAID; and indeed, USAID has initiated climate change oriented programs for Vietnam. Notably, USAID has identified Vietnam as a priority country in this

---

[37] U.S. Forest Service, *Climate Change in Vietnam: Assessment of Issues and Options for USAID Funding* (Washington, D.C.: USFS, February 2011), 39-40.

sector, and its efforts to date have focused on clean energy, adaptation and sustainable land use.[38]

Despite its seeming incongruity, the USFS involvement in such programs highlights the cross-cutting aspects of climate change, and a requirement for multiple – sometimes unexpected – contributors to the effort. Expertise developed by the USFS in addressing climate change domestically[39] can prove particularly suitable in providing international assessment and program design. Similarly, the U.S. Geological Survey (another Department of the Interior agency) is collaborating with Vietnam through the Delta Research and Global Observation Network, exploring how rising sea levels are likely to impact the Mekong delta.[40] Introducing yet another stream of activity, the Department of Commerce's National Oceanic and Atmospheric Administration (NOAA) signed a 2001 Memorandum of Understanding with Vietnam's National Hydro-Meteorological Service (discussed above) to provide advanced weather models and expertise to build capacity for prediction, warning and management of storm-induced flooding.[41] Subsequent to that agreement, joint activities have included events specifically looking at the impacts of climate variability (and also introducing the U.S. Office of Foreign Disaster Assistance into the discourse).[42] In reviewing these interwoven efforts by multiple branches of the U.S. government, the absence of the Department of Defense (DOD) is noteworthy. While U.S. Pacific Command (USPACOM) engages in many regional exercises, exchanges and training (including disaster management),

---

[38] USAID, "Global Climate Change: Vietnam," *USAID*, 25 October 2013, accessed 27 October 2013, http://www.usaid.gov/vietnam.
[39] U.S. Forest Service, "Climate Change Home," *USFS*, 16 September 2013, accessed 22 September 2013, http://www.fs.fed.us.
[40] USFS, *Climate Change*, 31.
[41] NOAA, "NOAA's Weather Service signs agreement to improve Vietnamese river, flood forecasts," *NOAA*, 9 January 2001, accessed 3 October 2013, http://www.noaanews.noaa.gov.
[42] National Weather Service, "Joint U.S.-Vietnam Climate Training Workshop," *NWS,* June 2009, accessed 3 October 2013, http://www.nws.noaa.gov.

an explicit DOD climate change platform on par with the agencies described above is still developing. However, U.S. national assessments certainly view the topic as a national security issue that merits an interagency approach.[43] Moreover, discussion at an August 2013 international military forum hosted by USPACOM and the Australian Department of Defense, which brought together many of the nations described above as at risk for climate change consequences (including Vietnam), suggests there is significant regional appetite for a military component to consequence management.[44]

Counter-Arguments

But perhaps DOD is wise to move slowly into the fray. Climate change, particularly on the scale that can threaten the consequences outlined here, remains a forecast rather than an established fact. The most cursory web-search for climate change news will readily produce a wide range of dissenting critiques; and even within the climate change consensus, the forecasts can produce a wide range of potential impacts over various time scales (exacerbated by the paucity of detailed local data and models noted earlier). In this light, assistance undertaken to address climate change may be, at best, a waste of resources; and at worst, actually harmful by slowing economic growth. However, such an argument is flawed on two counts. First, it merits note that much of the debate stems more from U.S. domestic political rivalries than it does from dissent within the scientific community;[45] and allowing such domestic divisions to limit assistance along lines explicitly requested by partner nations

---

[43] Defense Science Board, *Trends and Implications of Climate Change for National and International Security*, U.S. Government White Paper (Washington, D.C.: Office of the Under Secretary of Defense for Acquisition, Technology and Logistics, October 2011), xiv.

[44] Raymond Tusi, "Pacific Forum Highlights Military Resisilience to Climate Change," *USPACOM*, 22 April 2013, accessed 27 October 2013, http://www.pacom.mil/media/news/2013/04/22.

[45] Nate Silver, *The Signal and the Noise* (New York: Penguin, 2012), 409-410.

undermines efforts to establish the trust and influence which inspires many of our international engagement and assistance efforts. Moreover, as described above, climate change remediation programs follow two tracks – mitigation and adaptation. While mitigation efforts (e.g. emission controls) might arguably have unnecessary or harmful impacts to economic modernization; a whole range of adaptation measures (e.g. improved water resource management or disaster response) are "no regrets" strategies. That is, they produce a net public benefit regardless of whether, or to what degree, climate change progresses – a positive effect that is amplified by the regional nature of the issue.

However, this distinction can in turn give rise to another objection. Specifically, if issues like water management and disaster preparedness do provide a net public benefit, and are already being addressed by traditional assistance mechanisms – is there value added by packaging them under a new "climate change" banner? Worse, could enthusiasm for the new rubric undermine or unnecessarily complicate extant successful programs? While duly recognizing the risk associated with bureaucracies that become overly focused on the "next big thing," this argument fails to acknowledge the genuine impediments that emerge from multiple independent assistance efforts, especially in cross-cutting sectors. The U.S. inter-agency community is widely cognizant of those impediments, and actively pursues program frameworks designed to overcome them by encouraging (or requiring) multiple agencies and departments to work under unified rubrics where appropriate. Climate change provides precisely such a framework to harness and synchronize a wide variety of independent actors both within the U.S. Government and among international partners, as highlighted in the assistance review above. Moreover, such an approach fits well with increased Congressional tendency to encourage/compel inter-agency programmatic cooperation through appropriation of "dual-key" funding mechanisms like 1206 and the Global Security Contingency Fund (GSCF).[46]

Conclusion

The response to these objections reinforces two main themes raised in the introduction to this paper, namely: the cross-cutting nature of the problem that requires coordination of diverse actors to respond, and the regional nature of the challenge. Likewise, returning to the introduction, it should now be clear that both elements of the opening thesis are valid. First, Vietnam is indeed vulnerable to significant climate change consequences, as demonstrated both by the raw science and by the variety of local and international efforts already being assembled to address these consequences. Second, the nature and scope of these remediation efforts frames the utility of expanded U.S. assistance in this sector by meeting a number of important screening criteria. In initially considering such an effort, it is vital that both donor and recipient nations have a common conception of the problem, manifested here in both Vietnamese and U.S. technical assessments and policy documents. Next, the recipient country must actively desire assistance in the given sector, a desire Vietnam has positively and widely expressed. To make such support effective, local partners must be available, a role fulfilled in this context by MNRE and NHS; while international partners can enhance that effectiveness, a factor amply in evidence in this case. Finally, spoilers should be actively avoided – in the Vietnamese context the most likely such actor is China; but Beijing's relative neglect of this topic actually provides an opportunity.

---

[46] Andrew Shapiro, "A New Era for U.S. Security Assistance," *Washington Quarterly* 35.4 (Fall 2012): 33.

Recommendations

Consequently, this paper recommends that the U.S. government sustain and expand its support to Vietnam, with an increasing effort to harness a variety of activities under the climate change framework. Of particular utility might be pursuing climate change thematic Congressional appropriations that include dual-key authorities (like GSCF), and inter-agency/regional structures (like the Global Peacekeeping Operations Initiative). This regional approach could usefully by applied across programs in multiple countries, either through independent bi-lateral arrangements or extant multi-lateral mechanisms (like the Japan-supported ASEA disaster network). "No regret" initiatives are likely the most appropriate (and least contentious in a domestic political context); and especially in regards to disaster management, could provide an expanded role for DOD in a sector less prone to aggravating Chinese sensibilities than more conventional security affairs.

Adopting such an approach will provide critical support to Vietnam as it faces the storm clouds of climate change, and may allow it (like Conrad's sailors) to find that *"They may say what they like, but the heaviest seas run with the wind. Facing it – always facing it – that's the way to get through."*[47]

---

[47] Conrad, *Typhoon*, 178.

BIBLIOGRAPHY

Adger, W., P. Mick Kelly and Nguyen Huu Ninh, eds. *Living with Environmental Change: Social vulnerability, adaptation and resilience in Vietnam.* New York: Routledge, 2001.

Bruun, Ole, and Thorkill Casse, eds. *On the Frontiers of Climate and Environmental Change: Vulnerabilities and Adaptations in Central Vietnam.* New York: Springer, 2013.

Conrad, Joseph. *Typhoon.* New York: Putnam, 1902.

Defense Science Board Task Force. *Trends and Implications of Climate Change for National and International Security.* U.S. Government White Paper. Washington, D.C.: Office of the Undersecretary of Defense for Acquisition, Technology and Logistics, October 2011.

Dillon, C.P., and Mark Andrews. *1997 Annual Tropical Cyclone Report.* Pearl Harbor, HI: Joint Typhoon Warning Center, 1997.

Field, Christopher B., Vincent Barros, Thomas F. Stocker, Qin Dahe, et al. *Managing the Risks of Extreme Events and Disasters to Advance Climate Change Adaptation.* New York: Cambridge University Press, 2012.

Institute of Strategy and Policy on Natural Resources and Environment, Government of Vietnam. "Viet Nam Assessment Report on Climate Change." *United Nations Environment Program*, 2009. Accessed 10 September 2013, http://www.unep.org/climatechange/adaptation/Scienceand Assessments/VietnamAssessmentReport/tabid/29576/Default.aspx.

Intergovernmental Panel on Climate Change. "Climate Change 2013: The Physical Science Basis*." IPCC*, September 2013. Accessed 10 October 2013. http://www.ipcc.ch/ar5/wg1/#.UnMJghaYdz8

Intergovernmental Panel on Climate Change. *Managing the Risks of Extreme Events and Disasters to Advance Climate Change Adaptation.* New York: Cambridge, 2012.

Liotta, P.H. and Allan W. Shearer. *Gaia's Revenge.* Westport, CT: Praeger, 2007.

Ministry of Natural Resources and Environment, Government of Vietnam. "Climate Change, Sea Level Rise Scenarios for Vietnam." *Preventionweb*, June 2008. Accessed 10 September 2013. http://www.preventionweb.net/files/11348_ClimateChangeSeaLevelScenariosforVi.pdf.

Ministry of Natural Resources and Environment, Government of Vietnam. "Home page." *Government Portal*. Accessed 15 September 2013. http://noccop.org.vn/modules.php?name=Airvariable_Public$menuid=62.

Prime Minister's Office, Government of Vietnam. "National Strategy on Climate Change." *Government Portal*, 5 December 2011. Accessed 15 September 2013. http://www.chinhphu.vn/portal/page/portal/English/strategiesdetails?categoryid=30&articleid=10051283

Shapiro, Andrew. "A New Era for U.S. Security Assistance," *Washington Quarterly* 35.4 (Fall 2012): 23-35.

Silver, Nate. *The Signal and the Noise.* New York: Penguin, 2012.

Sterling, Eleanor Jane, Martha Maud Hurley and Le Duc Minh. *Vietnam: A Natural History*. New Haven: Yale University Press, 2006.

U.S. Forest Service. *Climate Change in Vietnam: Assessment of Issues and Options for USAID Funding*. U.S. Government White Paper. Washington, D.C.: USFS, February 2011.

www.ingramcontent.com/pod-product-compliance
Lightning Source LLC
Chambersburg PA
CBHW080631180526
45168CB00007B/3129